I0503287

The Implicit Price of Aquatic Grasses

Dennis Guignet, Charles Griffiths,
Heather Klemick and Patrick Walsh

Working Paper Series

Working Paper # 14-06
December, 2014

The Implicit Price of Aquatic Grasses

By:

Dennis Guignet*, Charles Griffiths, Heather Klemick, and Patrick Walsh

National Center for Environmental Economics
US Environmental Protection Agency

Last Revised: December 3, 2014

*Corresponding Author
  National Center for Environmental Economics
  US Environmental Protection Agency
  Mail Code 1809 T
  1200 Pennsylvania Avenue, N.W.
  Washington, DC 20460
  guignet.dennis@epa.gov

The views expressed in this paper are those of the author(s) and do not necessarily represent those of the U.S. Environmental Protection Agency. In addition, although the research described in this paper may have been funded entirely or in part by the U.S. Environmental Protection Agency, it has not been subjected to the Agency's required peer and policy review. No official Agency endorsement should be inferred. We thank Lisa Wainger and participants at the Northeastern Agricultural and Resource Economic Association's 2014 Meetings and Resources for the Future's Academic Seminar Series for helpful comments.

The Implicit Price of Aquatic Grasses

By:
Dennis Guignet, Charles Griffiths, Heather Klemick, and Patrick Walsh
December 2014

Abstract:

Almost 30% of aquatic grasses worldwide are either lost or degraded (Barbier et al, 2011). The Chesapeake Bay is no exception, with levels of submerged aquatic vegetation (SAV) remaining below half of the historic levels. This decline is largely attributed to excessive nutrient and sediment loads degrading Bay water quality. SAV provide many important functions to natural ecosystems, many of which are directly beneficial to local residents.

To understand the implicit value residents place on SAV and the localized ecosystem services it provides, we undertake a hedonic property value study using residential transaction data from 1996 to 2008 in eleven Maryland counties adjacent to the Chesapeake Bay. These data are matched to high resolution maps of Baywide SAV coverage. We pose a quasi-experimental comparison and examine how the price of homes near and on the waterfront vary with the presence of SAV. On average, waterfront and near-waterfront homes within 200 meters of the shore sell at a 5% to 6% premium when SAV are present. Applying these estimates to the 185,000 acre SAV attainment goal yields total property value gains on the order of $300 to 400 million.

JEL Classification: Q51 (Valuation of Environmental Effects); Q53 (Air Pollution; Water Pollution; Noise; Hazardous Waste; Solid Waste; Recycling)

Keywords: aquatic grasses; Chesapeake Bay; ecological input; ecosystem services; hedonics; submerged aquatic vegetation; SAV; water quality

I. INTRODUCTION

Almost 30% of aquatic grasses worldwide are either lost or degraded (Barbier et al, 2011). The Chesapeake Bay in the United States is no exception, with levels of submerged aquatic vegetation (SAV) remaining far below historic levels. The Chesapeake Bay is perhaps the largest estuary in North America and third largest in the world (Malmquist, 2009; CBP, 2012; NOAA, 2014; UVA, 2014), making it a vital natural amenity that provides numerous services to society and broader ecological systems. Based on historic levels of SAV, the Chesapeake Bay Program (CBP) and its state and federal partners have set a goal of achieving 185,000 acres of submerged aquatic vegetation (SAV) in the Chesapeake Bay and its tidal tributaries (CBP, 2014). Although the amount of SAV fluctuates from year to year, total acreage has continued to be well below half of this goal. Existing SAV can be damaged directly by human activities, such as boating, dredging, beach alterations, and aquaculture. Nutrient and sediment loads also degrade water quality and block essential sunlight from reaching aquatic plants. Excessive sedimentation further hampers growth by burying existing plants.

SAV provide many important functions to natural ecosystems, including food and habitat for wildlife, nutrient sequestration, and increased dissolved oxygen levels. SAV also serve as a good indicator of overall water quality because they are sensitive to both improvements and declines in water quality. Further, SAV help deter erosion and dissipate wave energy, which can be directly beneficial to local residents and users of the Bay. At the same time, SAV could be seen as a disamenity by swimmers, some boaters, and those participating in other recreational activities. The value society implicitly places on SAV as an input to the production of these various ecological services and amenities has yet to be reliably estimated in the literature (Barbier et al., 2011).

To better understand the net value local residents place on SAV and the localized services it provides, we undertake a hedonic property value study using residential transaction data from 1996 to 2008 in eleven Maryland counties adjacent to the Chesapeake Bay and its tidal tributaries. These data were matched to high spatial resolution data on Baywide SAV coverage. Taking advantage of spatial and temporal variation in the presence of SAV, we pose a quasi-experimental comparison and examine how the price of waterfront and non-waterfront homes in close proximity to the Bay varies with the presence of SAV. This study is one of only a few nonmarket valuation studies of SAV. In fact, to our knowledge this is the first hedonic property value study focusing on SAV in a tidal estuary, and where the SAV are largely composed of native grasses that may be viewed as a net amenity.

In the next section we provide some background on aquatic grasses in the Chesapeake Bay, and argue why local residents may value SAV, directly or indirectly. We then review the related nonmarket valuation literature and the unique contributions of this study in section III. In section IV we outline the hedonic property value model and in section V discuss the data. The results of the empirical analysis are presented in section VI, followed by concluding remarks in section VII.

II. BACKGROUND

II.A. Aquatic Grasses in the Chesapeake Bay

Going back to at least the 1930s, the Chesapeake Bay has historically supported about 185,000 acres of SAV (CBP, 2014). The decline in SAV density and coverage was first evident in the 1960s, and further accelerated in the 1970s (Kemp et al, 2005; Orth and Moore, 1983). Nutrient enrichment and excessive sediment loads entering the Bay substantially contributed to SAV reductions (Kemp et al., 2005; Orth and Moore, 1983). Nutrient enrichment and subsequent eutrophication block necessary sunlight from reaching these aquatic plants. Increased sediment deposition further reduces water clarity and can bury young plants.

From 2007 to 2010, CBP (2014) reports that over $6.2 million was used to fund the restoration, monitoring, and assessment/research of SAV in the Chesapeake Bay ($2.6 million of which was in Maryland). Numerous planting efforts have taken place throughout the Bay, where seeds or seedlings are dispersed across the Bay floor. However, these efforts often fail to produce beds that persist for more than a few years, with poor water quality being a key factor (Shafer and Bergstrom, 2008; Kemp et al., 2005).[1] Over the last several decades there have been extensive efforts at the local, state, and federal levels to reduce nutrient and sediment loads entering the Chesapeake Bay, and ultimately to improve water quality. This includes President Obama's 2009 Executive Order 13508, which led to the establishment of Total Maximum Daily Loads (TMDLs) to limit the amounts of nitrogen, phosphorous, and sediment entering the Bay.

Although SAV have recovered somewhat in parts of the upper Bay and other areas, many regions still remain devoid of SAV (Kemp et al., 2005). As shown in Figure 1, Baywide SAV acreage has remained largely below 45% of the historic 185,000 acres. SAV levels fluctuate from year to year due to climatic events, such as hurricanes and tropical storms (Kemp et al., 2005), but as of 2012 (the most recent year for which complete SAV data are available) the Bay remains at only 26% of the historic SAV levels. Preliminary estimates suggest that SAV increased slightly in 2013, but this is still only at 32% of the historic levels.[2]

---

[1] In contrast, in coastal bays adjacent to the Chesapeake, where water quality and turbidity are within a tolerable range for SAV, Orth et al. (2012) found that additional seeds led to rapid expansion of the SAV.
[2] This recent expansion in 2013 is largely due to rapid increases of widgeongrass; a species known for boom and bust cycles. Given concerns of the lack of SAV species diversity in these beds, it has yet to be seen whether this recent improvement will last in the longer-term (Blankenship, 2014).

II.B. Why Local Residents May Care about SAV

Although the notion of the environment as an input to the "production" of various goods and services has been around for a while (Lynne et al., 1981), this concept was most recently formalized by Boyd and Krupnick (2013), who define the ecological production function and discuss how features of the natural environment can be viewed as an ecological endpoint, input, or both. Ecological endpoints are outputs from the ecological "production" process, and are features of the environment that people directly care about, and that therefore directly enter a household's utility function. In contrast, ecological inputs are features of the environment that only affect household utility indirectly, in that a change in inputs may affect the quantity or quality of the resulting ecological service or amenity a household "consumes". Aquatic grasses may be viewed as an ecological input, endpoint, or both. As an endpoint, it is possible that some households may view the presence or quality of SAV as a direct amenity or disamenity. At the same time SAV may also be an ecological input, meaning that residents may not value the presence of SAV in itself, but they do inherently value SAV for its contribution in "producing" environmental commodities they do care about.

In the Chesapeake Bay SAV grow in all salinity regimes, and include a variety of species. The most common species are Eelgrass, Widgeon Grass, Wild Celery, Hydrilla, Redhead Grass, Sago Pondweed, and Eurasian Milfoil (Orth et al., 2013). SAV are typically submerged plants, with foliage growing at or near the water surface, implying that SAV can be visible from the shore. Some species have a simple grass-like structure (e.g., Eelgrass), whereas others have more complex structures and can form sparse to dense mats of foliage at the water surface (e.g., Hydrilla)

(MD DNR, 2010). Similar to plants that grow on land in this region, SAV undergo seasonal cycles. SAV growth begins in the spring and beds reach their peak density around the summer months. Senescence begins in late fall and coverage is sparse during the winter (Orth et al., 2012; Hansen and Reidenbach, 2013), suggesting that SAV may not be visible in the winter months.[3] SAV contribute to a variety of ecological services and amenities that local residents may value, and therefore may in turn be capitalized into local housing values. For example, residents may enjoy watching or hunting waterfowl and other local wildlife, or partaking in recreational activities such as fishing and crabbing at the waters near their home. SAV provide a critical habitat, food source, and predator protection for many ecologically and economically important species, including blue crab and juvenile fish (Barbier et al., 2011; Kemp et al., 2005), as well as waterfowl (Johnston et al., 2002; MD DNR, 2010). The role of SAV as a nursery for juvenile fish and shellfish, contributing to species density, individual growth, juvenile survival, and movement to adult habitat, is also often cited in the literature (Heck et al., 2003).

Local residents may also value higher levels of water quality and clarity. SAV contribute to the ecological production of water quality and clarity through several mechanisms. First, aquatic plants produce oxygen through photosynthesis, which in turn increases dissolved oxygen levels in the water and better supports aquatic life (NOAA, 2008; Kemp et al., 2005). Second, SAV beds filter excess nutrients in the water column (Barbier et al., 2011), which in turn decreases the frequency of algae bloom events and hypoxia. Kemp et al. (2005) show that if SAV beds in the Upper Chesapeake Bay were restored to historic levels, it would remove about 45% of nitrogen loads entering the Upper Bay. They argue that even partial restoration of SAV would "substantially

---

[3] Senescence in this context refers to the die off of foliage in the winter months.

help mitigate effects of nutrient loading" (pg 13). Third, SAV attenuate wave energy, which slows water flows and filters sediment out of the water column (Chen et al., 2007). This wave attenuation and the binding of sediments on the bay floor by SAV roots and rhizomes also deter the re-suspension of sediment into the water column (Ward et al., 1984). This particle trapping and binding of sediment by SAV increases water clarity, and further encourages photosynthesis and nutrient assimilation (Kemp et al., 2005).

The wave attenuation, sediment deposition, and binding of deposited sediment by SAV also contribute to coastal protection and erosion control (Barbier et al., 2011). As waves move towards the coast wave energy is diminished by SAV leaves. The resulting coastal protection is highest when SAV beds are dense and occupy the entire water column (Chen et al., 2007; Koch et al., 2009; Ward et al., 1984). Wave attenuation increases sediment deposition, which leads to shallower waters and further contributes to wave attenuation (Koch et al., 2009).

Although local residents may value SAV directly or as an input in producing services and amenities, it is also possible that SAV could be viewed as a disamenity by some, particularly recreationists (Kragt and Bennett, 2011). For example, swimmers may prefer relatively clear un-vegetated waters where they can "see their feet" (EPA, 2013), and do not have to worry about stepping on or swimming through vegetation. Recreational boaters and perhaps some fishermen may dislike SAV because the plants can get caught on their fishing lines or propellers. While some of the resulting ecological endpoints SAV contribute to may be relatively widespread (such as improvements in water clarity/quality, or increased fish and shellfish populations due to the nursery effect), others are very local in nature, including: coastal protection and reduced erosion; improved clarity from decreased sediment suspension; increased presence of fish, shellfish, and waterfowl due to the role of SAV as habitat; as well as some nuisance effects. With these localized endpoints in mind, the purpose of this hedonic property value study is to examine the net welfare impact SAV have on residents living on, or in close proximity to, the Chesapeake Bay waterfront.

III. LITERATURE REVIEW

In a recent review, Barbier et al. (2011) report finding few reliable estimates of the value of SAV. They note a few studies attempting to monetize the value of SAV as an input in commercial fisheries, but beyond that, estimates are sparse. Among the few studies Barbier et al. identify, the focus largely entails ecological simulation models that append a unit value to the simulated change in biomass based on commercial market prices (Watson et al., 1993; McArthur and Boland, 2006, Sanchirico and Mumby, 2009).

A few other studies have relied on statistical relationships or used bioeconomic simulation modeling to examine the value of SAV. Kahn and Kemp (1985) relate SAV abundance to fish stock and catch in the Chesapeake Bay, and account for both commercial and recreational fishing values in their welfare calculations. Johnston et al. (2002) simulate how changes in eelgrass in the Peconic Bay percolate through the ecological system and ultimately affect fish, shellfish, and waterfowl populations. Assigning unit values based on commercial prices and recreational viewing and hunting values, Johnston et al. estimate an asset value of $17,759 per acre of eelgrass (2010$).[4] The corresponding value to create a new acre of eelgrass is estimated at $9,996. Focusing on the Puget Sound, Plummer et al. (2013) conduct a similar exercise and append commercial and

---

[4] All dollar estimates converted to USD 2010$ based on the US Bureau of Labor Statistics' Annual US Average "All Urban Consumers – Consumer Price Index (CPI); http://www.bls.gov/cpi/cpid1404.pdf, Table 12 (accessed June 18, 2014).

recreational fish values to projected increases in fish populations.[5] A key advantage of such simulation models is that they capture the underlying biological structure of the ecosystem. At the same time, the main drawback is that they rely heavily on professional judgement and do not allow tests for statistical significance of the estimated values (Johnston et al., 2002).

Non-market valuation approaches, such as stated preference (SP) methods, on the other hand, do allow for tests of statistical significance. In a parallel SP study of the Peconic Bay, Johnston et al. find that local residents value an acre of eelgrass at $0.12 per year per household, which summed over all 73,423 households in the Peconic Bay Estuary System translates to an annual value of $8,589 per acre (Johnston et al, 2001, 2002). In the George's Bay Estuary in Tasmania, Kragt and Bennett (2011) found that the median household has a WTP of $0.02 to $0.04 for an additional acre of seagrass. The authors do, however, question the use of seagrass beds as an indicator for measuring public preferences for estuary health; highlighting the disparity between the science and the general public's understanding.

This is a concern with SP surveys valuing SAV, and other ecological inputs, in general. The ecological production function relating such inputs to the ecosystem services and amenities people value must be clearly and quantitatively communicated. Otherwise survey respondents rely on subjective beliefs about how inputs influence the "production" of the endpoints they care about. Such beliefs are unknown to the researcher and can be wildly unfounded, thus bringing into question the validity of the resulting welfare estimates (Boyd and Krupnick, 2013; Johnston et al., 2013).

---

[5] Plummer et al. (2013) also report simulated changes in bird and whale populations due to SAV, but note the lack of any monetary unit value to apply to these changes.

A key advantage of revealed preference methods, such as the hedonic approach, is that we need only observe the initial ecological input (i.e., SAV) and the end result of how household behavior is influenced. The underlying ecological production function does not need to be modeled or communicated, as is the case with bioeconomic simulation and SP studies. The hedonic property value approach allows for statistical inference and gets directly at the monetized outcome of interest. Of course by sidestepping the underlying ecological production processes the approach is unable to quantify changes in intermediate inputs and the ecological endpoints themselves, which may be of great interest to stakeholders. Further, even though detailed knowledge is not required, it is still crucial to have the underlying endpoints and processes in mind when making causal claims of the effect of the ecological inputs (SAV in this case) on property values.

There are a few previous hedonic studies analyzing the impacts of a specific type of SAV, Eurasian Milfoil. These studies find that waterfront property values around freshwater lakes depreciate with increased Milfoil (Halstead et al., 2003; Horsch and Lewis, 2009; Zhang and Boyle, 2010; Tuttle and Heintzelman, 2014). As discussed in section II, SAV can pose both desirable and undesirable features to households. Milfoil is an invasive species that floats on the water surface, can spread rapidly from lake to lake, and is often considered a disamenity because it reduces the quality of recreational activities (e.g., swimming, boating, and fishing). Milfoil can also accelerate eutrophication and have uncertain irreversible effects. Our study is unique because we focus on an iconic coastal estuary where the aquatic grasses are mainly native species, and offer several services and amenities that local residents may value.[6]

------

[6] Among the seven most common species of SAV in the Chesapeake Bay and tidal waters, only two are non-native species to the Bay (Orth et al., 2013; MD DNR, 2010). The first is Eurasian Milfoil, which although fairly invasive, has died back and stabilized in the Bay since the 1960s. Even though the species is invasive, in the Bay it still offers local ecological services (e.g., habitat for juvenile fish and crabs). The second invasive species is Hydrilla, which

Even though there have been no hedonic property value studies on SAV as a potential amenity, there have been several hedonic property value studies on some of the ecological endpoints that SAV help provide. For example, studies generally find a price premium for wider beaches and lower risks of erosion (Landry et al., 2003; Landry and Hindsley, 2011), but this may not always be the case (Ranson, 2012). Numerous studies also report that houses near clearer or better quality waters, all else constant, are valued at a premium.[7] There have also been several hedonic studies of wetlands, which offer similar ecological services as SAV. McConnell and Walls (2005) review the nonmarket valuation literature on open space, including hedonic studies on the impacts of wetlands on nearby home values. They find that home price impacts tend to vary depending on whether the wetland is in an urban or rural area.

Three hedonic studies have previously examined water quality in the Chesapeake Bay. Leggett and Bockstael (2000) analyze the effect of fecal coliform concentrations on waterfront home values in Anne Arundel County, Maryland, and find that higher concentrations resulted in a statistically significant decrease in waterfront home prices. Poor et al. (2007) analyze the impact of ambient water quality on homes in another Maryland county (St. Mary's), and find a significant negative correlation between concentrations of dissolved inorganic nitrogen and total suspended solids and property values. Most recently, Walsh et al. (2014) conduct a hedonic analysis in 14

---

can also be beneficial, particularly in areas generally devoid of native SAV, by providing ecological services that would not otherwise be there (e.g., fish habitat and food source for waterfowl). However, Hydrilla can grow aggressively and overcome native SAV. Hydrilla can also be considered a nuisance because its dense beds can impede recreation in waterways, particularly along the Potomac River (MD DNR, 2010; CBF 2014).

[7] The majority of these hedonic are of homes that are on or near the waterfront of freshwater lakes; particularly those in the Northeast US (Young, 1984; Michael et al., 2000; Boyle et al., 1999; Boyle and Taylor, 2001; Poor et al., 2001, Gibbs et al., 2002) and Florida (Walsh et al., 2011a, 2011b, Bin and Czajkowski, 2013). Water clarity, as measured by secchi depth, is the most commonly used measure of water quality. Other measures have been used, including concentrations of fecal coliform, total nitrogen or phosphorous, chlorophyll *a*, and total suspended solids, among others (Epp and Al-Ani, 1979; Poor et al., 2001; Leggett and Bockstael, 2001; Walsh et al., 2011a). Identifying the appropriate measures of water quality remains the focus of much research (Griffiths et al., 2012).

Maryland counties bordering the Chesapeake Bay and tidal waters. Although they find heterogeneity in the implicit price of light attenuation (which is inversely related to water clarity), their subsequent meta-analysis reveals a statistically significant average elasticity of 0.06% for waterfront homes and 0.01% for non-waterfront homes up to 500 meters away, suggesting that local residents do hold a premium for clearer waters, all else constant (Klemick et al., 2014). We use this same dataset and extend these earlier works with the aim of estimating the implicit price local residents place on SAV.

IV. EMPIRICAL MODEL

We estimate multiple hedonic property value regression models, where the dependent variable is the natural log of the transaction price for home i in neighborhood j, when it was sold in period t (p_{ijt}). The hedonic price is estimated as a function of characteristics of the housing structure itself (e.g., interior square footage, number of bathrooms), as well as of the parcel (e.g., lot acreage) and its location (e.g., distance to major roads, presence in a floodplain), which we denote as x_{ijt}. The price of a home also depends on overall trends in the housing market, which are accounted for by annual and quarterly dummy variables (M_t). Lastly, we posit that the presence of submerged aquatic vegetation (SAV) may affect local housing values. SAV is measured using an indicator variable denoting the presence of SAV (SAV_{ijt}) interacted with a vector of dummy variables denoting whether a home is within various proximity buffers from the waterfront (W_{ij}). The equation to be estimated is:

$$\ln p_{ijt} = x_{ijt}\beta + M_t\alpha + W_{ij}\theta + (W_{ij} \times SAV_{ijt})\gamma + v_j + \varepsilon_{ijt} \qquad (1)$$

where ε_{ijt} is a normally distributed error term. The coefficients to be estimated are $\boldsymbol{\beta}$, $\boldsymbol{\alpha}$, $\boldsymbol{\theta}$, v_j, and of particular interest, $\boldsymbol{\gamma}$.

In most specifications we allow for neighborhood specific fixed effects (v_j), which absorb all time invariant influences on property values within a particular locale j. We vary the spatial scale of these fixed effects across our regression models (e.g., census tract or block group, as defined by the 2000 U.S. Census). In our preferred models these fixed effects are at the "block group-bay buffer" level, which denote the spatial intersection between block groups and a buffer of 0 to 500 meters from the Bay. Therefore, in our preferred specifications all time invariant unobserved factors associated with the waters and neighborhood among waterfront and near waterfront homes within a particular block group are controlled for, including otherwise unobserved factors that might be correlated with the presence of SAV.

The ultimate objective is to estimate the implicit price of SAV, conditional on all other characteristics of the home and its location, including proximity to the waterfront, which is captured by $\boldsymbol{\theta}$. The coefficient $\boldsymbol{\gamma}$ can usually be interpreted as a semi-elasticity, but since we measure SAV using binary indicators, following Halvorsen and Palmquist (1980) we calculate the percent change in value due to the presence of an SAV bed as:

$$\% \Delta \boldsymbol{p} = 100 \times (e^{\gamma} - 1) \tag{2}$$

In the preferred "block group-bay buffer" (BG-BB) fixed effects models, this can be interpreted as the change in property value due to SAV, relative to other waterfront or near waterfront homes within that specific block group, all else constant.

The presence of SAV beds vary spatially along the waterfront of a block group and over time, thus facilitating a quasi-experiment where we compare the value of two waterfront (or near waterfront) homes: one where SAV are present the year of sale (the "treated" group) and the other where SAV are not present (the "control" group). Consider waterfront homes in the example depicted in Figure 2 for a few block groups in Anne Arundel County, MD. The different color land areas denote the BG-BB neighborhood fixed effects. In 1996, notice there are only a few waterfront homes adjacent to SAV beds. These homes are considered the "treated" group in our quasi-experimental framework. We isolate the price premium associated with such homes, relative to other waterfront homes that are sold within that same BG-BB area. Additionally, we take advantage of temporal variation in SAV beds (both gains and losses) within a single block group, allowing for a spatial difference-in-difference approach (Horsch and Lewis, 2009). As seen in Figure 2, several new SAV beds arose between 1996 and 2000 in this section of the Chesapeake Bay.

Under this difference-in-difference framework, %Δp from equation (2) can be interpreted as the average treatment effect. This price differential reflects the net effect of all changes in *localized* endpoints due to the presence of SAV. Since the control group consists of other homes within the same waterfront neighborhood, to the extent desirable and undesirable features of SAV spillover to neighboring properties, the quasi-experimental comparison will be confounded. Thus, the capitalization effects estimated in this analysis capture only the net effect of SAV and the amenities and nuisances it provides that are very local in nature (see section II.B).

14

V. THE DATA

We focus on 11 Maryland counties that are adjacent to the Chesapeake Bay and its tidal tributaries (see Figure 3).[8] We focus on Maryland because a dataset of residential transactions is compiled annually and these data are formatted in a similar fashion across counties, which facilitates cross-county comparisons and allows us to pool the transactions and estimate a single Baywide hedonic regression.[9] We focus on arms-length transactions from 1996 to 2008 of single-family homes and townhomes within four kilometers of the Bay and its tidal tributaries. To minimize the influence of outliers attention is restricted to homes where the real price was between $40,000 and $4,000,000 (2010$), and where the parcel size was less than or equal to 100 acres (leaving 199,833 sales). Finally, we eliminated transactions of waterfront or non-waterfront homes within 500 meters of the waterfront where SAV data were missing the year of sale, leaving a final sample size of n=195,373 transactions.[10]

These data are accompanied by a wealth of variables describing each home, the date a home is sold, and the amount it actually sold for, which is the dependent variable in the hedonic price regressions. The geographic coordinates of each residential parcel are also included, allowing

[8] Montgomery and Wicomico Counties were excluded because there were no transactions observed where SAV beds were present. Baltimore City, Calvert, and Somerset Counties were disregarded due to very few observed sales where SAV were present.

[9] Data were obtained from Maryland Property View, which is a compilation of the tax assessment and transaction databases across all Maryland counties.

[10] In some years SAV data could not be collected in certain portions of the Bay due to weather conditions and excessive turbidity. Such cases were identified based on documentation in the VIMs annual reports (e.g., Orth et al., 2013). SAV data were considered missing if aerial photography and SAV mapping data were stated as not being available for a particular Bay segment in a given year. The 4,459 transactions with missing SAV data were mainly in 2001 in Baltimore and Anne Arundel Counties. The regression results discussed in Section VI are robust if instead of excluding these transactions, the SAV variables are coded to zero and a companion missing dummy variable included.

us to calculate the distance of each parcel to the waterfront, and identify whether SAV beds are present along that specific part of the waterfront.

V.A. Submerged Aquatic Vegetation Beds

The Virginia Institute of Marine Science (VIMS) collects and maintains spatially explicit annual data on SAV coverage throughout the Chesapeake Bay and its tidal tributaries.[11] These data are based primarily on aerial photographs taken during numerous flights over a several-month period each year between May and December. The aerial photographs are interpreted and validated through comparisons to ground surveys, ultimately producing annual high-resolution geographic information systems (GIS) data of SAV beds throughout the Chesapeake Bay.[12] These data efforts go back until at least the mid-1980s. We focus on the SAV datasets from 1996 to 2008, which coincides with our data on residential property transactions.

Each residential parcel in the study area is matched to the nearest SAV bed as of the year of sale. We compare the distance to that SAV bed with a parcel's distance to the waterfront, and create an indicator variable (*SAV*) equal to one if the SAV bed distance is less than or equal to the distance to the waterfront plus a 50 meter buffer. In other words, *SAV* is equal to one if there is an SAV bed within 50 meters of the shoreline near each home. *SAV* is then interacted with dummy variables denoting whether a home is within a certain distance buffer from the waterfront. The first interaction term *waterfront* × *SAV* equals one for homes that are on the waterfront and where an SAV bed is within that distance plus a 50 meter buffer. Similarly, the interaction terms *water 0-*

---

[11] VIMS, http://web.vims.edu/bio/sav/index.html, accessed April 17, 2014.
[12] Further details are documented in each annual report (see, for example, Orth et al., 2013).

200 m × SAV and *water 200-500 m × SAV* denote non-waterfront homes that are within 0 to 200 meters and 200 to 500 meters of the waterfront, respectively, and where an SAV bed is within the distance to the waterfront plus a 50 meter buffer. A buffer of 50 meters was chosen to approximate for the presence of SAV along the shoreline.

Table 1 shows descriptive statistics for the dummy variables denoting waterfront homes, as well as non-waterfront homes that are within 0 to 200 meters or 200 to 500 meters of the Bay tidal waters. About 5% of our sample of home transactions are on the waterfront, and approximately 14% and 19% of sales are of non-waterfront homes within the 0 to 200 meter and 200 to 500 meter bay proximity buffers, respectively. Considering the entire sample, only 0.7% of transactions were of waterfront homes where SAV was present the year of the sale. Similarly, only about 1.4% and 2.1% of sales correspond to non-waterfront homes within the 0 to 200 meter and 200 to 500 meter buffers, respectively, and where SAV were present.

The number of residential transactions within each water proximity buffer and where SAV were present (SAV=1) or not (SAV=0) is shown by county in Table 2. Conditional on all other observables, one can think of these sales as the "treated" and "control" groups, respectively, in our quasi-experimental comparison to identify the effect of SAV on home values at different distances from the Bay.

V.B. Housing Bundle Characteristics

The transaction data from Maryland Property View (MDPV) includes numerous variables denoting various features of a home and parcel. Descriptive statistics for the transaction price (p_{ijt}) and home structure characteristics that are included in x_{ijt} are displayed in Panel A of Table

3. The mean price across all transactions in the sample is $238,507 (median is $217,696). To serve as a proxy for overall quality and features of all structures, we include the assessed value for all structures as an explanatory variable in the hedonic regressions.[13] Among transactions where the assessed value for all improvements was available, we find an average of about $109,665. The average home in the sample is just under 30 years in age at the time of sale, has an interior size of 1,432 square feet, a parcel size of 0.58 acres, and about 1.5 bathrooms. About 20% of the sample are townhomes, as opposed to single-family homes.

Table 3.B and 3.C display descriptive statistics for the various location-oriented variables that are included as explanatory variables in x_{ijt}. Table 3.B includes relatively local spatial characteristics, which may vary among homes within the same locale (namely the same Census block group). For example, based on land use data from MDPV we derive binary indicator variables denoting parcels that are located in high- or medium-density residential areas. Distances from each parcel were also calculated to various local amenities and disamenities (e.g., nearest primary road, urban area, beach). Table 3.C includes broader spatial characteristics that may not vary much within block groups, including distances to the nearest wastewater treatment plant, major city, and power plant.[14] Based on the parcel coordinates we also identify the Census tract and block group where each parcel is located. To proxy for surrounding land uses, we link each parcel to the proportion of its neighborhood (as defined by the 2000 Census block groups) devoted

---

[13] Leggett and Bockstael (2000) used this same approach to account for overall features of the home structure but not the location, since assessed land value is a separate variable.

[14] Primary road GIS data were obtained from ESRI's North American Street Map. The 27 wastewater treatment plants were selected from all Major NPDES (National Pollutant Discharge Elimination System) permitted facilities within five kilometers of the Bay tidal waters. These data were obtained from EPA's Federal Registry System (http://www.epa.gov/enviro/html/frs_demo/geospatial_data/geo_data_state_combined.html, accessed July 10, 2013). Primary or major cities were defined as those with populations greater than 250,000 according to ESRI's USA Major Cities shapefile. Urbanized areas and clusters, as defined in the 2010 U.S. Census Bureau's "Urban Areas" gazetteer file, are used to represent secondary and tertiary cities (or business districts).

to various land uses delineated by MDPV (industrial, urban, agriculture, beach, etc). The census tracts and block groups were also used to define spatial (or neighborhood) fixed effects, as discussed below.

V.C. Addressing Location Specific Confounding Factors

The presence of SAV throughout the Bay is not random. SAV require specific conditions in order to grow and thrive. In some cases these requirements could be correlated with other factors making coastal areas in some parts of the Bay more or less desirable, which could in turn affect property values. Great care is taken to control for such factors and minimize any potential for omitted variable bias.

For example, SAV cannot grow as well in waters where there is a lot of wave energy and strong currents because the seeds do not have the opportunity to fully settle and root themselves to the bay floor (Shafer and Bergstrom, 2008). Further, such waters may kick up and carry lots of sediment, burying SAV and deterring growth. At the same time, areas with heavy waves may be more susceptible to erosion, storm surges, and flooding, which in turn could be capitalized into home values. To account for these factors we used floodplain maps developed by FEMA to create an indicator variable denoting whether a home is located within a 100 year floodplain. Such areas include coastal hazard zones which are susceptible to additional hazards associated with waves induced by storms. As shown in Table 3.B, about 5.3% of the home sales in our sample are within a floodplain.

Water depth is also accounted for in the hedonic regressions by including an indicator variable denoting depths between 0 to 2 meters. In the Chesapeake Bay SAV historically grow in areas where the water depth is between 0 to 2 meters (Kemp et al., 2005). At the same time deeper waters may be more desirable to local residents because it allows for different recreational activities, such as boating and the ability to have a dock onsite. Data of water depth were obtained from the National Oceanic and Atmospheric Administration's (NOAA) digital elevation models of estuarine bathymetry, which has a fairly high spatial resolution of 90 meters squared.[15] Each residential parcel was matched to the water depth at the nearest portion of the Bay or tidal waters. According to this local water depth measure, about 98% of sales were of homes with a water depth of 2 meters or less. The average depth is 0.52 meters. Focusing on just waterfront homes or non-waterfront homes within 500 meters, about 96.3% and 96.6% of the home transactions correspond to water depths of two meters or less.

Sunlight is another critical component for SAV growth. Relatively clear waters allow more light to reach SAV, thus promoting growth. At the same time, local residents likely value water clarity, and as shown throughout the hedonic literature, these values are reflected in the housing market (e.g., Boyle and Taylor, 2001; Poor et al., 2001; Gibbs et al., 2002; Walsh et al., 2011a, 2011b). As a robustness check, in some of our hedonic regressions we control for local water clarity, as measured by the mean spring and summertime light attenuation (denoted as KD) during or just prior to the time of sale.[16] These data were obtained from EPA's Chesapeake Bay Program (CBP). Monthly water quality measurements are taken from numerous monitoring stations and are then interpolated to grid cells with a maximum size of 1km×1km, and that cover the entire

---

[15] National Oceanic and Atmospheric Administration (NOAA), National Ocean Service (NOS) Estuarine Bathymetry, http://estuarinebathymetry.noaa.gov/bathy_htmls/M130.html, accessed July 17, 2014. In earlier drafts we included an alternative depth measure provided by EPA's Chesapeake Bay Program (CBP). CBP's depth measures are provided at a spatial resolution of about 1 square kilometer. The SAV results discussed in section VI are robust to the inclusion of this alternative broader water depth measure, or both.

[16] Light attenuation can be converted to secchi disk measurement (SDM) in meters based on the following statistical relationship: $K_D = 1.45/SDM$ (EPA 2003).

mainstem of the Bay and its tidal tributaries (see Walsh et al., 2014 for details). Among waterfront homes and non-waterfront homes within 500 meters of the bay, mean KD is 2.34 (min=0.62 and max=9.66). This corresponds to a secchi depth of approximately 0.62 meters (ranging from 0.15 to 2.36 meters).

Lastly, we include geographic fixed effects to absorb all time invariant influences on property values within a particular locale. The inclusion of these fixed effects, along with controlling for various potentially confounding factors directly, facilitate a cleaner quasi-experiment for identifying the implicit price of SAV.

VI. RESULTS

VI.A. Main Hedonic Regression Results.

In the main results we pool transactions across all 11 counties and estimate a single hedonic regression. We include interaction terms between annual time and individual county dummies in order to allow broader housing market trends to vary by county. The results for several specifications are reported in Table 4. Model 4.A is the simplest (and most restrictive) model. Although we allow county specific constant terms and time trends, all other coefficients corresponding to the home and location attributes are constrained to be the same across counties. Specifications 4.B and 4.C impose similar restrictions, but the spatial fixed effects are more refined. Model 4.B includes census tract (CT) fixed effects and 4.C allows for block group (BG) fixed effects.

Only the coefficient estimates of interest are presented in table 4, but all attributes shown in table 3 are included as explanatory variables.[17] Comparison of specifications 4.A through 4.C reveal that the property value changes associated with waterfront proximity and SAV are robust (with the point estimates declining slightly as the spatial fixed effects become more refined). The estimates corresponding to *waterfront* suggest that homes located on the waterfront sell for a hefty premium compared to homes located at distances greater than 500 meters from the Bay, all else constant. Non-waterfront homes located 0 to 200 meters and 200 to 500 meters from the Bay also sell for a premium, although as one may expect this premium is smaller, and in the 200 to 500 meter buffer is statistically indistinguishable from zero (at least when census tract or block group level fixed effects are included).

Of most interest are the estimates corresponding to the interaction terms between the bay proximity buffers and the presence of SAV. In models 4.A through 4.C, the coefficient estimates for these interaction terms are fairly robust. For example, plugging the coefficient estimates corresponding to *waterfront × SAV* into equation 2 suggests that waterfront homes that have SAV along the shoreline tend to sell for an additional 4.7% to 7.9% premium, compared to waterfront homes where SAV are not present. We see a slightly higher premium associated with non-waterfront homes within 0 to 200 meters (*water 0-200 × SAV*), ranging from 6.8% to 8.0%. The premiums associated with SAV are not statistically different across the waterfront and 0-200 meter buffers.[18] The premium associated with SAV for homes 200 to 500 meters from the waterfront (*water 200-500 × SAV*) range from 1.7% to 2.3% but are statistically insignificant.

---

[17] The full regression results for models 4.A through 4.D are provided in Appendix A.
[18] Nonlinear Wald tests fail to reject the null hypothesis that that the percent change in price corresponding to SAV are statistically equal across the waterfront and 0-200 meter buffers (p-value <=0.05 for models 4.A through 4.C).

Model 4.D in Table 4 includes our preferred block group-bay buffer (BG-BB) fixed effects. The BG-BB fixed effects account for all time invariant price differences associated with each waterfront neighborhood, and therefore provide the cleanest quasi-experimental comparison. The BG-BB fixed effects are defined by splitting all block groups within 500 meters of the Bay into two separate fixed effects, one for homes in that block group that are within 500 meters of the waterfront, and another denoting all other homes in that block group. Note that the water 200-500 meter coefficient is omitted from 4.D and subsequent models because this is the omitted category in the BG-BB fixed effect specifications (essentially this coefficient is allowed to vary freely across block groups). The SAV coefficients are similar to the previous specifications. Again following equation (2), the results suggest that the presence of SAV leads to a 5.0% premium among waterfront homes. For non-waterfront homes within 0 to 200 meters and 200 to 500 meters we find a 6.7% and 2.3% premium associated with the presence of SAV (although the latter is only statistically significant at the 10% level).[19]

While it may be somewhat surprising that the SAV coefficient for the 0 to 200 meter buffer is higher than that on waterfront homes, these two coefficients are not statistically different from one another in any of the specifications. In addition, the gradient as one moves away from the waterfront is somewhat different when the price effects are converted to implicit prices. The mean price for homes where SAV are not present is $675,364 among waterfront homes, and $308,187 and $264,196 for non-waterfront homes within 0-200 and 200-500 meters, respectively. The mean

---

[19] As a robustness check, this model was re-estimated with additional interaction terms denoting whether SAV were present at any time during the study period. In doing so, we further control for confounding price effects associated with areas where SAV tend to grow in general. The estimated premiums for *waterfront × SAV* and *water 0-200 × SAV* are slightly smaller, but robust, suggesting a 4.5% and 4.2% premium, respectively, when SAV are present the year of the sale. This further supports that the implicit price estimates are capturing the effects of SAV, and not just other spatially correlated unobservables.

implicit prices of SAV are estimated by multiplying these average prices by the estimated percent changes in property values, implying that the presence of SAV is associated with a $33,968 premium among waterfront homes and a $20,566 premium among non-waterfront homes within 0 to 200 meters (both are statistically significant with p-values < 0.01). We find a smaller $6,115 premium among homes that are in the 200 to 500 meter buffer, but this is only marginally significant (p=0.088).

The implicit price estimates do suggest a decreasing price gradient associated with SAV; however, the implicit price estimates are not statistically different between the waterfront and 0 to 200 meter buffers. The implicit prices are statistically different between the 0 to 200 and 200 to 500 meter buffers (p=0.0204). Although we find some evidence of a decreasing price gradient, it is important to note that such a relationship may not necessarily hold depending on the various desirable and undesirable features of SAV, and how these features affect households at different distances from the water. For example, all households may enjoy increased bird and wildlife watching associated with SAV, but undesirable effects of SAV on swimming or boating may have a more adverse impact on waterfront households compared to others, which on net could suggest a non-monotonic price gradient.

In any case, these estimates can be interpreted as premiums relative to other waterfront or near waterfront homes within the same waterfront neighborhood (as defined by the BG-BB fixed effects). The counterfactual in this quasi-experiment are other waterfront or near waterfront homes in that same block group where SAV were not present the year of sale.

Although all 11 counties are in the Chesapeake Bay region it is unclear whether pooling the data as we have done thus far is appropriate, at least statistically speaking. These counties, or subsets of these counties, could be considered separate housing markets. Numerous interaction terms are included in model 4.E to allow all coefficients to vary by county. Only the coefficients corresponding to the SAV interaction terms are constrained to be the same across counties. This provides a clean Baywide average estimate of γ, while still allowing for market heterogeneity across counties. Likelihood ratio tests clearly reject the null model (4.D) in support of this more flexible specification ($p < 0.0000$).

The coefficient estimates corresponding to SAV remain robust. Plugging these estimates into equation (2) yields an estimated 5.7% premium among waterfront homes and 6.4% among homes in the 0-200 meter buffer. The corresponding mean implicit price estimates are $38,611 and $19,676, respectively. The results suggest that SAV have a small and statistically insignificant effect on homes beyond 200 meters.

VI.B. Additional Hedonic Regressions and Robustness Checks

The hedonic regressions in Table 5 are estimated using only the 74,594 sales of homes within 500 meters of the Bay tidal waters. Focusing on this more homogenous set of homes along the waterfront facilitates an even cleaner quasi-experimental comparison between waterfront and near waterfront homes where SAV are, and are not, present. Model 5.A is the same as 4.E, but only includes homes within 500 meters of the tidal waters. The results corresponding to the SAV interactions are almost identical. Although not reported here, we also estimate variants of model 5.A that include SAV dummies equal to 1 if SAV were present during the last 3 years. The results were similar, and even suggest that these longer sustaining SAV beds are associated with a slightly higher 7.0% premium among waterfront homes and an 11.1% premium among homes within 0-200 meters. Again we find that these two estimates are not statistically different from each other.

The results are also robust to the inclusion of local water clarity in model 5.B, as measured by the natural log of light attenuation (*ln(KD)*). Clearer waters can be capitalized in property values (Boyle and Taylor, 2001; Poor et al., 2001; Gibbs et al., 2002; Walsh et al., 2011a, 2011b), while at the same time could be correlated with the presence of SAV. Including *ln(KD)* further reduces the potential for any omitted variable bias. The SAV estimates are robust, and it is reassuring that the light attenuation coefficients are of the expected sign and significance. KD is inversely related to secchi disk measurement (*SDM*) following the approximate statistical relationship: $K_D = 1.45/SDM$ (EPA 2003), where *SDM* is measured in meters. A negative sign implies a premium for clearer waters. In fact, the coefficients corresponding to *ln(KD)* are elasticities, and so a 1% improvement in clarity would suggest 0.09% increase in waterfront home values, and a 0.04% increase among non-waterfront homes within 0 to 200 meters of the bay.[20]

In model 5.C we distinguish between SAV beds of different vegetation densities. Based on visual inspection, the VIMs datasets categorize SAV beds according to a density scale of 1 to 4, where 1 = very sparse (<10% coverage); 2 = sparse (10-40%); 3 = moderate (40-70%); and 4 = dense (70-100%) (Orth et al., 2013). Dummy variables are created denoting which density category a SAV bed falls within, and these are then interacted with the bay buffer dummy variables. Depending on how SAV density translates into the ecological services, amenities, and

---

[20] These estimates are slightly larger than the average results across counties reported by Walsh et al. (2014), although their analysis differs from the current study because it includes three additional counties, estimates an econometric model with spatial lag and autocorrelation terms, and does not include spatial fixed effects.

disamenities, that local residents care about, one may expect the implicit price of SAVs to vary with density.

Focusing first on waterfront homes, we see that the SAV density coefficients across the first three density categories are statistically significant and are fairly similar to each other and to the previous models, ranging from 0.0571 to 0.0679. The slightly smaller and statistically insignificant 0.0312 coefficient corresponding to the densest SAV category (*waterfront × SAV density 4*) may suggest that the premium associated with SAV is less when the vegetation is too dense; perhaps because it deters recreational activities available to waterfront households. In any case, an F-test shows that the estimates across density categories for waterfront homes are not statistically different from each other (p=0.5849).

For homes in the 0 to 200 meter buffer, the point estimates are all similar, falling within 0.0507 and 0.0635. F-tests clearly fail to reject the null hypothesis that these estimates are statistically equivalent (p=0.9111). Perhaps the amenities and ecological services SAV provide to these non-waterfront households are fairly similar across density categories. Lastly, in agreement with the previous models, there is no evidence that SAV have a statistically significant impact on homes beyond 200 meters from the waterfront.

We next examine county heterogeneity by re-estimating variants of the BG-BB fixed effects model separately for each county. The results are presented in Appendix B, but in short suggest some heterogeneity in the premiums associated with SAV. Whether this heterogeneity reflects differences across counties in terms of household preferences, housing supply, or inherent features of SAV and related amenities and services (as well as nuisances) remains uncertain. It is also possible that these results reflect the fact that some counties have very few home sales with

SAV in the various buffer zones (see Table 2), and so caution is warranted in interpreting these county specific results.

Considering the SAV coefficient estimates for waterfront homes across all 11 counties, nine of the coefficient estimates are positive, although only three are statistically significant (p-value < 0.05). In only one county (Charles) is a statistically significant negative estimate found. This is particularly interesting because in Charles County we also find positive and statistically significant coefficients for *water 0-200 m × SAV* and *water 200-500 m × SAV*. Perhaps SAV are particularly bothersome to waterfront residents in Charles County, but are still viewed as a net amenity among non-waterfront residents. In the 0 to 200 meter buffer we find positive coefficients in 8 of the 11 counties, but the coefficients are significant in only two of these counties (p-value < 0.05). In the 200 to 500 meter buffer we find positive SAV coefficients in 6 of the 11 counties, but again the estimates are only significant in two of these counties. In both the 0-200 and 200-500 meter buffers we find no significant negative coefficients on the SAV interactions across all 11 counties.

VI.B. Chesapeake Bay 185,000 Acre SAV Goal

Based on historic record and photographic evidence, it is estimated that the Chesapeake Bay has historically supported about 185,000 acres of SAV (CBP, 2014). As a result, the Chesapeake Bay Program and its state and federal partners have used that acreage as a goal for the Bay and its tidal tributaries. Using the estimates from the hedonic analysis, we illustrate what the benefits of achieving this goal might be, at least in terms of local property values.

In order to estimate the incremental price impacts of the SAV goal, an appropriate baseline must be determined. Given the numerous unknowns regarding future land use, population growth, best management practices, and how such things translate into changes in SAV, we refrain from making any future baseline projections. Instead, we take the 85,914 acre SAV coverage in 2009 and assume this as our baseline. We do so for three reasons: (i) SAV levels are relatively high that year (see figure 1), so our estimates can be considered conservative in that sense; (ii) this is the most recent year we have SAV GIS data spatially linked to residential parcels; and (iii) this is the most recent year for which we have assessed values for each residential property in the study area.

Figure 4 displays the spatial coverage of SAV in the baseline year and under the attainment goal. GIS data on the SAV goal coverage were obtained from the Chesapeake Bay Program.[21] In Maryland, gains in SAV are particularly noticeable along the Eastern Shore, Anne Arundel County, and the mouths of the Potomac and Patuxent Rivers. The baseline and attainment goal SAV data were spatially linked to all town- and single-family homes in the 11 Maryland counties. Among the 83,729 homes that are waterfront or within 200 meters of Bay, 10,736 have SAV present along the nearby shoreline in the 2009 baseline (see Table 6). This almost doubles under the SAV attainment goal, reaching 20,955 homes.

Two different approaches are taken to estimate the total change in property values. In the first, we simply multiply the net change in homes with SAV in the waterfront and 0-200 meter buffer, by the corresponding mean implicit price estimate from Model 4.E. As shown in table 7, this yields a total change in property values of about $326 million. The second approach is more refined and spatially explicit in that it accounts for the gain or loss in SAV and assessed value at

----

[21] Personal Communication, Chesapeake Bay Program, April 30, 2014.

each individual home. Multiplying the assessed value by $\%\Delta\boldsymbol{p}$ estimated from model 4.E, as appropriate for each individual home and the corresponding change in SAV, and then summing the gains and losses in value over all parcels, yields a total change in property values of about $398 million. Although the difference in these estimates is noticeable, the 95% confidence intervals largely overlap, and it is reassuring that the more sophisticated second approach yields similar results to the first, fairly simple, approach.

VII. CONCLUSION

Aquatic grasses often play a key role in aquatic ecosystems, and thus provide a plethora of ecological services and amenities that society values. At the same time almost 30% of aquatic grasses worldwide are either lost or degraded (Barbier et al, 2011). This study focuses on one of the largest estuaries in the world, the Chesapeake Bay, where submerged aquatic vegetation (SAV) have remained far below historic levels.

Focusing on eleven Maryland counties adjacent to the Bay and its tidal waters, we employ hedonic property value methods to estimate the net value local residents place on SAV beds along the shoreline. Hedonic methods are particularly advantageous in valuing SAV because many of the ecological services and amenities SAV provide are local in nature, such as: coastal protection and reduced erosion, increased wildlife for recreational purposes, and improved water clarity. Although there have been a few hedonic studies on the adverse property value impacts from invasive aquatic vegetation (Halstead et al., 2003; Horsch and Lewis, 2009; Zhang and Boyle, 2010; Tuttle and Heintzelman, 2014), to our knowledge this is the first hedonic study on mainly native aquatic grasses in a large, iconic, coastal estuary, and where the aquatic grasses in question offer several services and amenities that local residents may value.

SAV can be thought of as an input in the production of ecological services and amenities (and sometimes disamenities) that directly enter households' utility functions. Examining how property values vary with this ecological input is advantageous in that changes in many ecological endpoints can be valued at once, while at the same time circumventing the need for complex ecosystem simulation models and the inherent assumptions behind them.

We utilize a quasi-experimental study design that relies on spatial and temporal variation in SAV and uses spatially refined "block group-bay buffer" fixed effects to control for all time invariant price influences associated with each individual waterfront neighborhood. We believe that the analysis provides credible evidence that homes tend to sell at a premium when SAV are present, and that this is suggestive of a causal relationship.

Our preferred specification suggests that, on average, waterfront homes where SAV are present sell at a 5.7% premium relative to other waterfront homes within the same waterfront neighborhood (but where SAV are not present). Similarly, non-waterfront homes within 200 meters of the bay sell at a 6.4% premium. These estimates translate to a mean implicit price of $38,611 and $19,676 per home, respectively, and are robust across numerous specifications. We find little evidence that SAV impact property values beyond 200 meters from the waterfront.

Applying these estimates to the 185,000 acre SAV attainment goal for the Chesapeake Bay, we find that, relative to an assumed 2009 baseline, this could lead to total property value gains on the order of $326 to $398 million. It is important to note that these estimates only reflect the localized impacts of SAV, and only to households on or near the waterfront in the eleven Maryland counties analyzed. SAV provide many ecological services that span a fairly broad geographic area, such as being a nursery for numerous ecologically and economically important species of fish and shellfish. The broader commercial, recreational, and nonuse values associated with SAV are not captured in this analysis. Incorporating such values through different market and nonmarket valuation methods is a valuable direction for future research on the value of aquatic grasses and other key ecological inputs, and the ecosystem services they provide.

WORKS CITED

Barbier, Edward B., Sally D. Hacker, Chris Kennedy, Evamaria W. Koch, Adrian C. Stier, and Brian R. Silliman (2011), "The value of estuarine and coastal ecosystems," *Ecological Monographs*, 81(2), 169-193.

Bin, O. and J. Czajkowski (2013). "The Impact of Technical and Non-technical Measures of Water Quality on Coastal Waterfront Property Values in South Florida." *Marine Resource Economics*, 28(1): 43-63.

Blankenship, Karl (2014), "SAV rebounds 24% in 2013," *Bay Journal*, April 26, 2014; http://www.bayjournal.com/article/sav_rebounds_24_in_2013, accessed June 18, 2014.

Boyd, James and Alan Krupnick (2013), "Using Ecological Production Theory to Define and Select Environmental Commodities for Nonmarket Valuation," *Agricultural and Resource Economic Review*, 42(1), 1-32.

Boyle, K. J. and L. O. Taylor (2001). "Does the Measurement of Property and Structural Characteristics Affect Estimated Implicit Prices for Environmental Amenities in a Hedonic Model." *Journal of Real Estate Finance and Economics,* 22(2/3): 303-318.

Boyle, K. J., P. J. Poor and L. O. Taylor (1999). "Estimating the Demand for Protecting Freshwater Lakes from Eutrophication." *American Journal of Agricultural Economics,* 81(5): 1118-1122.

CBF (Chesapeake Bay Foundation) (2014), "Invasive Plant Species", http://www.cbf.org/about-the-bay/chesapeake-bay/plants-of-the-chesapeake/invasive-species, accessed June 19, 2014.

CBP (Chesapeake Bay Program) (2012), https://www.chesapeakebay.net/discover, accessed April 18, 2014.

CBP (Chesapeake Bay Program) (2014), http://stat.chesapeakebay.net/, accessed June 19, 2014.

Chen, Shih-Nan, Lawrence P. Sanford, Evamaria W. Koch, Fengyan Shi, and Elizabeth W. North (2007), "A Nearshore Model to Investigate the Effects of Seagrass Bed Geometry on Wave Attenuation and Suspended Sediment Transport," *Estuaries and Coasts*, 30(2), 296-310.

EPA (2003), U.S. Environmental Protection Agency, "Ambient Water Quality Criteria for Dissolved Oxygen, Water Clarity and Chlorophyll a for the Chesapeake Bay and Its Tidal Tributaries", Office of Water. Annapolis MD, EPA 903-R-03-002.

EPA (2013), Environmental Protection Agency, "Additional Documents Available for Public Review Related to Willingness to Pay Survey for Chesapeake Bay Total Maximum Daily Load: Instrument, Pre-Test, and Implementation", Focus Group Report. 78 FR 38713, additional-documents-available-for-public-review-related-to-willingness-to-pay-survey-for-chesapeake, https://federalregister.gov/a/2013-15439, accessed April 22, 2014.

Epp, Donald, and K. S. Al-Ani (1979), "The Effect of Water Quality on Rural Nonfarm Residential Property Values," *American Journal of Agricultural Economics*, 61(3), 529-534.

Gibbs, J. P., J. M. Halstead and K. J. Boyle (2002), "An Hedonic Analysis of the Effects of Lake Water Clarity on New Hampshire Lakefront Properties," *Agricultural and Resource Economics Review* 31(1): 39-46.

Griffiths, C., H. Klemick, M. Massey, C. Moore, S. Newbold, D. Simpson, P. Walsh and W. Wheeler (2012). "U.S. Environmental Protection Agency Valuation of Surface Water Quality Improvements." *Review of Environmental Economics and Policy*.

Halstead, John, Jodi Michaud, and Shanna Hallas-Burt, and Julie Gibbs (2003), "Hedonic Analysis of Effects of a Nonnative Invader (Myriophyllum heterophyllum) on New Hampshire (USA) Lakefront Properties," *Environmental Management*, 32(3), 391-398.

Halvorsen, Robert, and Raymond Palmquist (1980), "The Interpretation of Dummy Variables in Semilogarithmic Equations," *The American Economic Review*, 70(3), 474-475.

Hansen, Jennifer C. R., and Matthew A. Reidenbach (2013), "Seasonal Growth and Senescence of Zostera marina Seagrass Meado Alters Wave-Dominated Flow and Sediment Suspension within a Coastal Bay," *Estuaries and Coasts*, 36, 1099-1114.

Heck, K. L. Jr., G. Hays, R. J. Orth (2003), "Critical evaluation of the nursery role hypothesis for seagrass meadows," *Marine Ecology Progress Series*, 253, 123-136.

Horsch, Eric J. and David J. Lewis (2009), "The Effects of Aquatic Invasive Species on Property Values: Evidence from a Quasi-Experiment," *Land Economics*, 85(3), 391-409.

Johnston, Robert J., Thomas A. Grigalunas, James J. Opaluch, Marisa Mazzotta, and Jerry Diamantedes (2002), "Valuing Estuarine Resource Services Using Economic and Ecological Models: The Peconic Estuary System Study," *Coastal Management*, 30, 47-65.

Johnston, Robert J., James J. Opaluch, Thomas A. Grigalunas, and Marisa J. Mazzotta (2001), "Estimating Amenity Benefits of Coastal Farmland," *Growth and Change*, 32, 305-322.

Johnston, Robert J., Eric T. Schultz, Kathleen Segerson, Elena Y. Besedin, and Mahesh Ramachandran (2013), "Stated Preferences for Intermediate versus Final Ecosystem Services: Disentangling Willingness to Pay for Omitted Outcomes," *Agricultural and Resource Economics Review*, 42(1), 98-118.

Kahn, James R. and W. Michael Kemp (1985), "Economic Losses Associated with the Degradation of an Ecosystem: The Case of Submerged Aquatic Vegetation in the Chesapeake Bay," *Journal of Environmental Economics and Management*, 12, 246-263.

Kemp, W. M., W. R. Boynton, J.E. Adolf, D. F. Boesch, W. C. Boicourt, G. Brush, J.C. Cornwell, T. R. Fisher, P. M. Gilbert, J. D. Hagy, L. W. Harding, E. D. Houde, D. G. Kimmel, W. D. Miller, R. I. E. Newell, M. R. Roman, E. M. Smith, J. C. Stevenson (2005), "Eutrophication of the Chesapeake Bay: historical trends and ecological interactions," *Marine Ecology Progress Series*, 303, 1-29.

Klemick, Heather, Charles Griffiths, Dennis Guignet, and Patrick Walsh (2014), "Explaining Variation in the Value of Water Quality Using Internal Meta-analysis," presented at the Northeast Agricultural and Resource Economics Association's Annual Meeting, Morgantown, WV, June 2014.

Koch, Evamaria W., Edward B. Barbier, Brian R. Silliman, Denise J. Reed, Gerardo ME Perillo, Sally D. Hacker, Elise F. Granek, Jurgenne H. Primavera, Nyawira Muthiga, Stephen Polasky, Benjamin S. Halpern, Christopher J. Kennedy, Carrie V. Kappel, and Erik Wolanksi (2009), "Non-linearity in ecosystem services: temporal and spatial variability in coastal protection," *Frontiers in Ecology and the Environment*, 7(1), 29-37.

Kragt, Marit E., and J.W. Bennett (2011), "Using choice experiments to value catchment and estuary health in Tasmania with individual preference heterogeneity," *The Australian Journal of Agricultural and Resource Economics*, 55, 159-179.

Landry, C. E. and P. Hindsley (2011), "Valuing Beach Quality with Hedonic Property Models." *Land Economics*, 87(1), 92-108.

Landry, C. E., A. G. Keeler and W. Kriesel (2003), "An Economic Evaluation of Beach Erosion Management Alternatives," *Marine Resource Economics,* 18(2), 105-127.

Leggett, C. G. and N. E. Bockstael (2000). "Evidence of the Effects of Water Quality on Residential Land Prices." *Journal of Environmental Economics and Management* 39(2): 121-144.

Lynne, Gary D., Patricia Conroy, and Frederick J. Prochaska (1981), "Economic Valuation of Marsh Areas for Marine Production Processes," *Journal of Environmental Economics and Management*, 8, 175-186.

Malmquist, David (2009), "How big is the Bay?", Virginia Institute of Marine Science, http://www.vims.edu/bayinfo/faqs/estuary_size.php, accessed April 18, 2014.

McArthur, Lynne C. and John W. Boland (2006), "The economic contribution of seagrass to secondary production in South Australia," *Ecological Modelling*, 196, 163-172.

McConnell, Virginia, and Margaret Walls (2005), "The Value of Open Space: Evidence from Studies of Nonmarket Benefits," Resources for the Future, Washington, D.C., January, 2005.

MD DNR (Maryland Department of Natural Resources) (2010), "Bay Grass Identification Key," http://www.dnr.state.md.us/bay/sav/key/complete_sav_key.pdf, accessed June 19, 2014.

Michael, H. J., K. J. Boyle and R. Bouchard (2000). "Does the Measurement of Environmental Quality Affect Implicit Prices Estimated from Hedonic Models?" *Land Economics* 76(2): 283-298.

NOAA (National Oceanic and Atmospheric Administration) (2014), "Where is the largest estuary in the United States?", http://oceanservice.noaa.gov/facts/chesapeake.html, accessed April 18, 2014.

NOAA (National Oceanic and Atmospheric Administration) (2008), "Dissolved Oxygen", http://oceanservice.noaa.gov/education/kits/estuaries/media/supp_estuar10d_disolvedox.html, accessed April 22, 2014.

Orth, Robert J., and Kenneth A. Moore (1983), "Chesapeake Bay: An Unprecedented Decline in Submerged Aquatic Vegetation," *Science*, 222(4619), 51-53.

Orth, Robert J., Kenneth A. Moore, Scott R. Marion, David J. Wilcox, and David B. Parrish (2012), "Seed addition facilitates eelgrass recovery in a coastal bay system," *Marine Ecology Progress Series*, 448, 177-195.

Orth, Robert J., David J. Wilcox, Jennifer R. Whiting, L. Nagey, Anna K. Kenne, and Erica R. Smith (2013), "2012 Distribution of Submerged Aquatic Vegetation in Chesapeake Bay and Coastal Bays," Special Scientific Report #155, Virginia Institute of Marine Science, College of William and Mary, Gloucester Point, VA; October, 2013.

Plummer, Mark L., Chris J. Harvey, Leif E. Anderson, Anne D. Guerry, and Mary H. Ruckelshaus (2013), "The Role of Eelgrass in Marine Community Interactions and Ecosystem Services: Results from Ecosystem-Scale Food Web Models," *Ecosystems*, 16, 237-251.

Poor, P. J., K. J. Boyle, L. O. Taylor and R. Bouchard (2001). "Objective versus Subjective Measures of Water Clarity in Hedonic Property Value Models." Land Economics 77(4): 482-493.

Poor, P. J., K. L. Pessagno and R. W. Paul (2007). "Exploring the hedonic value of ambient water quality: A local watershed-based study." *Ecological Economics,* 60(4): 797-806.

Ranson, Matthew (2012), "What Are the Welfare Costs of Shoreline Loss? Housing Market Evidence from a Discontinuity Matching Design," Discussion Paper 2012-07, Belfer Center for Science and International Affairs, Harvard University, Cambridge, MA; May 2012.

Sanchirico, James N., and Peter Mumby (2009), "Mapping ecosystem functions to the valuation of ecosystem services: implications of species – habitat associations for coastal land-use decisions," *Theoretical Ecology*, 2, 67-77.

Shafer, Deborah J. and Peter Bergstrom (2008), "Large-Scale Submerged Aquatic Vegetation Restoration in the Chesapeake Bay," US Army Corps of Engineers, ERDC/EL TR-08-02, Washington, D. C.; June, 2008.

Tuttle, Carrie, and Martin Heintzelman (2014), "A Loon on Every Lake: A hedonic Analysis of Lake Water Quality in the Adirondacks," Draft Paper, Clarkson University, Potsdam, NY, May.

UVA (University of Virginia) (2014), "The UVA Bay Game", http://www.virginia.edu/vpr/sustain/BayGame/thebay/, accessed April 18, 2014.

Walsh, P. J., J. W. Milon and D. O. Scrogin (2011a). The Property-Price Effects of Abating Nutrient Pollutants in Urban Housing Markets. Economic Incentives for Stormwater Control. H. Thurston, CRC Press: 127-145.

Walsh, P. J., J. W. Milon and D. O. Scrogin (2011b). "The Spatial Extent of Water Quality Benefits in Urban Housing Markets." Land Economics 87(4): 628-644.

Walsh, Patrick, Charles Griffiths, Dennis Guignet, and Heather Klemick (2014), "A Hedonic Analysis of Water Quality and the Chesapeake Bay TMDL," presented at the Northeast Agricultural and Resource Economics Association's Annual Meeting, Morgantown, WV, June 2014.

Ward, Larry G., W. Michael Kemp, and Walter R. Boynton (1984), "The Influence of Waves and Seagrass Communities on Suspended Particulates in an Estuarine Embayment," *Marine Geology*, 59, 85-103.

Watson, Reg A., Robert G. Coles, and Warren J. Lee Long (1993), "Simulation Estimates of annual Yield and Landed Value for Commercial Penaeid Prawns from a Tropical Seagrass Habitat, Northern Queensland, Australia," *Australian Journal of Marine and Freshwater Research*, 44(1), 211-219.

Young, Edwin (1984), "Perceived Water Quality and the Value of Seasonal Homes," *Journal of the American Water Resources Association*," 20(2), 163-166.

Zhang, Congwen, and Kevin Boyle (2010), "The effect of an aquatic invasive species (Eurasian watermilfoil) on lakefront property values," *Ecological Economics*, 70, 394-404.

FIGURES AND TABLES

Figure 1. Acreage of Aquatic Grasses in the Chesapeake Bay.

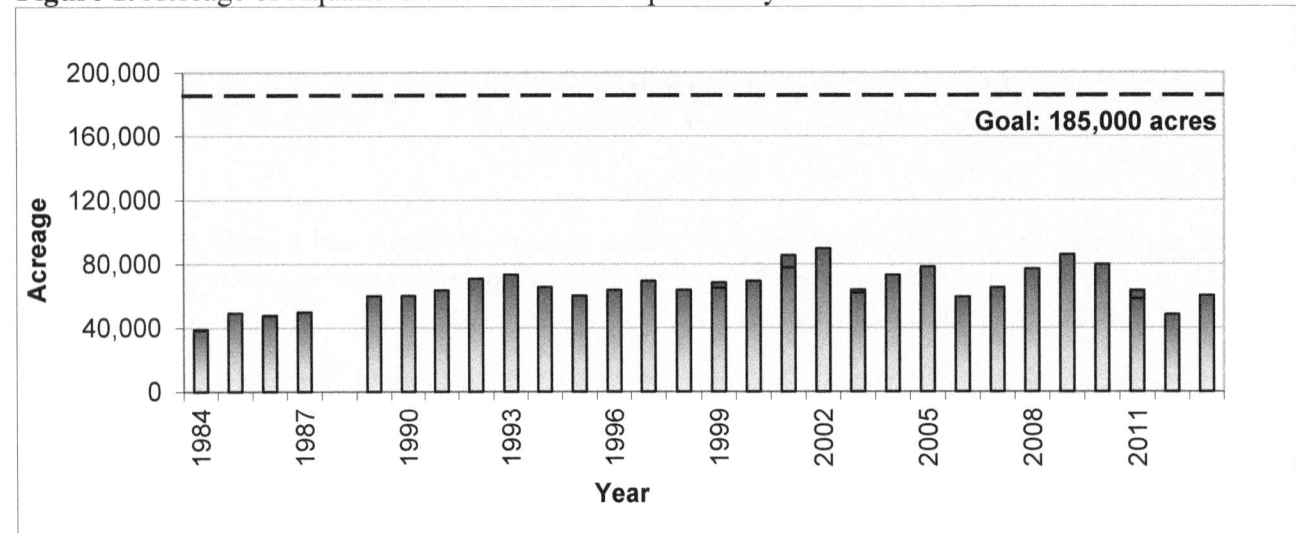

Source: Data of actual and estimated SAV acreage obtained from CBP (Chesapeake Bay Program), http://stat.chesapeakebay.net/, accessed December 2, 2014.

Note: In 1984, 1986, 1999, 2001, 2003, and 2011 SAV could not be surveyed in some segments of the Bay due to weather and excessive turbidity. In such cases SAV were estimated based on segment specific coverage in the previous year. In these six years the estimated SAV composed 1% to 10% of the total acreage in a given year (mean of 5%). See CBP website for details.

Figure 2. Example of block group by bay 500 meter buffer (BG-BB) fixed effects. Submerged aquatic vegetation (SAV) beds in portion of Anne Arundel County, MD.

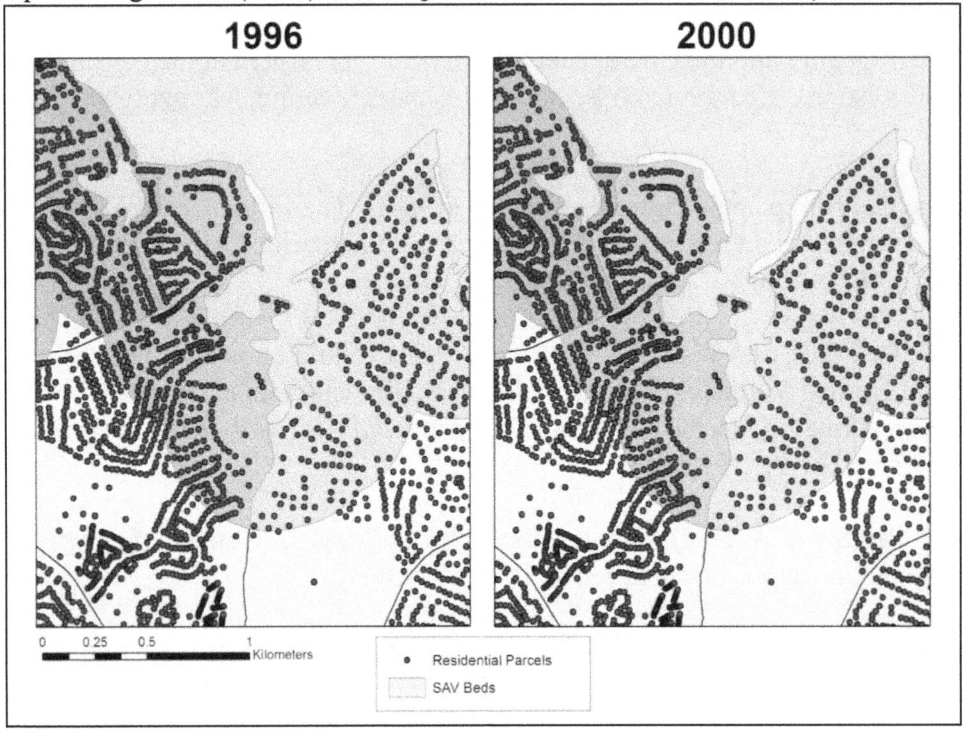

Figure 3. Study Area: Chesapeake Bay and Tidal Waters, and the 11 Maryland Counties featured in the hedonic property value analysis.

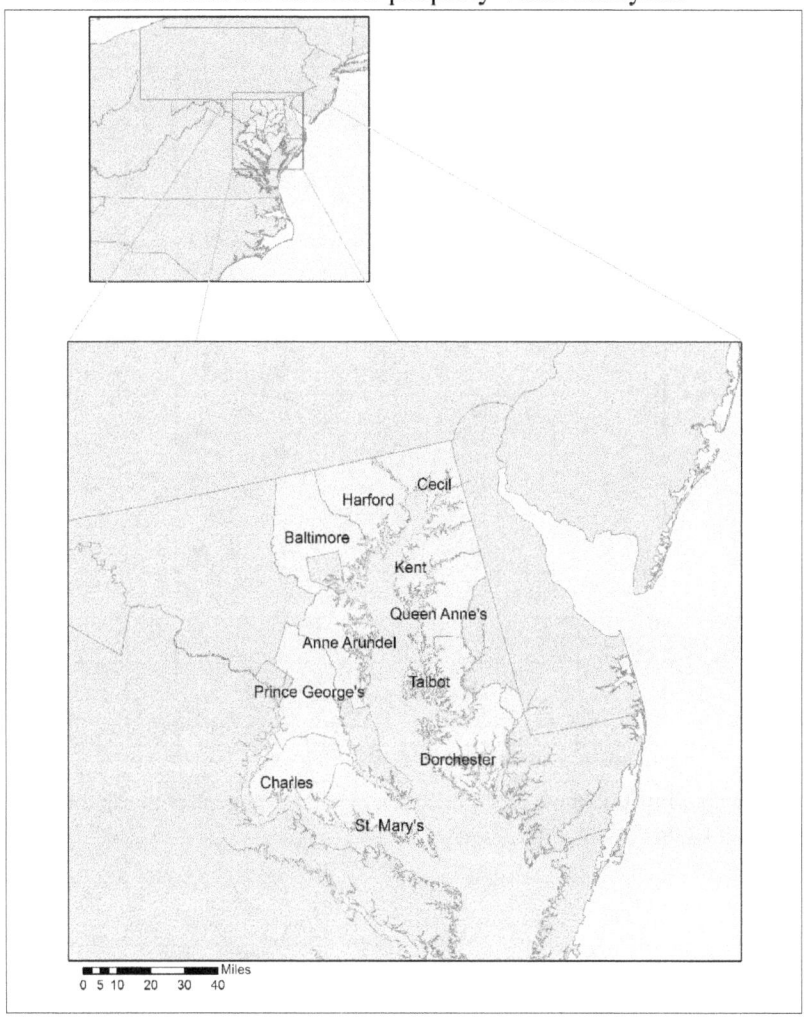

Figure 4. Submerged Aquatic Vegetation (SAV) Coverage: 2009 Baseline and 185,000 Acre SAV Goal.

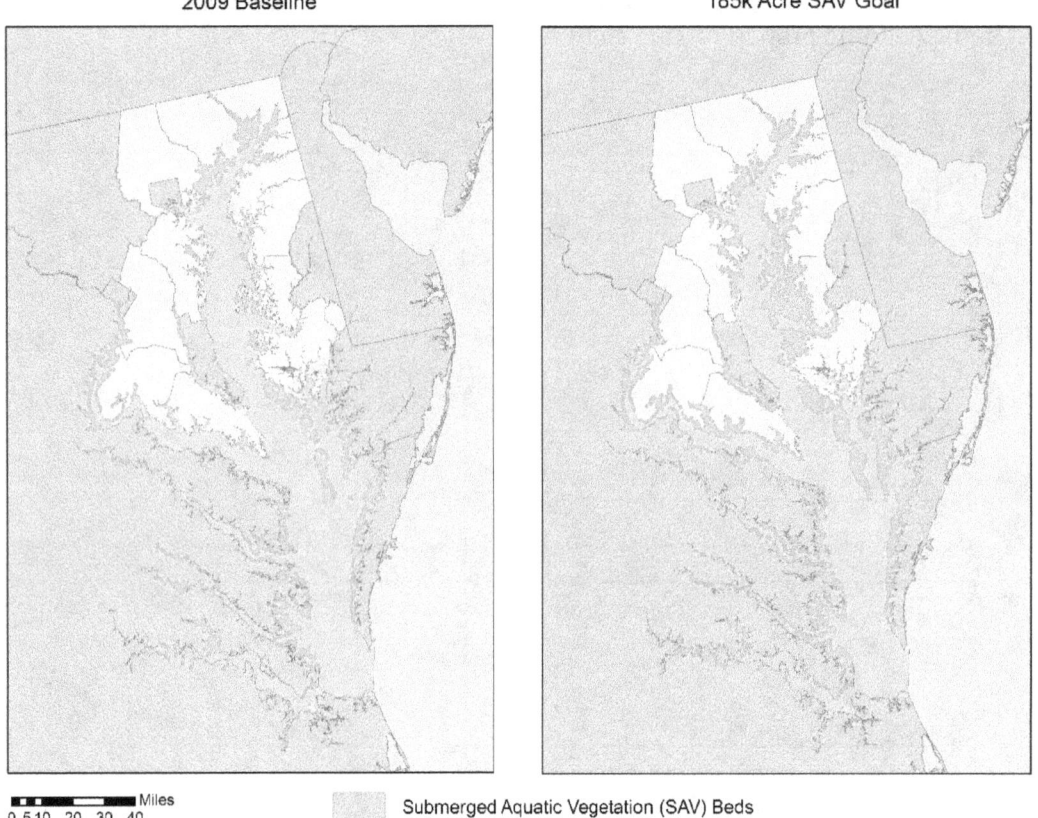

Note: SAV beds in both scenarios are drawn with a line thickness (2.00 line weight) that exaggerates the spatial coverage in order to facilitate visual comparison at this broad scale.

Table 1. Descriptive Statistics of Waterfront, Waterfront Proximity, and Submerged Aquatic Vegetation (SAV).

Variable[a]	Obs	Mean	Std. Dev.	Min	Max
Waterfront	195,373	0.050	0.217	0	1
× SAV	195,373	0.007	0.085	0	1
Non-waterfront: water w/in 0 to 200 meters	195,373	0.139	0.346	0	1
× SAV	195,373	0.014	0.117	0	1
Non-waterfront: water w/in 200 to 500 meters	195,373	0.193	0.395	0	1
× SAV	195,373	0.021	0.142	0	1

a.   All variables are dummy variables.

Table 2. Number of Residential Transactions with and without Submerged Aquatic Vegetation, by County and Waterfront Proximity Buffer.[a]

	Anne Arundel		Baltimore		Cecil	
	SAV=0	SAV=1	SAV=0	SAV=1	SAV=0	SAV=1
waterfront	3,172	280	1,035	236	247	126
0 to 200 m	12,600	624	5,084	642	1,057	352
200 to 500 m	16,527	921	6,899	613	1,156	462

	Charles		Dorchester		Harford	
	SAV=0	SAV=1	SAV=0	SAV=1	SAV=0	SAV=1
waterfront	136	115	502	139	162	27
0 to 200 m	250	208	533	35	952	189
200 to 500 m	355	356	1,042	38	1,354	750

	Kent		Prince George's		Queen Anne's	
	SAV=0	SAV=1	SAV=0	SAV=1	SAV=0	SAV=1
waterfront	347	36	60	31	905	170
0 to 200 m	454	47	593	97	1,270	262
200 to 500 m	752	111	1,645	275	1,979	259

	St. Mary's		Talbot	
	SAV=0	SAV=1	SAV=0	SAV=1
waterfront	520	19	1,170	246
0 to 200 m	567	47	1,145	214
200 to 500 m	723	41	1,247	189

a. Transactions of homes without SAV are denoted as SAV=0 and homes with SAV present are denoted as SAV=1.

Table 3. Home structure and location characteristics for sales within 4km of the waterfront (all 11 counties)

3.A. Home Structure and Parcel Characteristics

Variable	Obs	Mean	Std. Dev.	Min	Max
Dependent Variable: Transaction price[a]	195,373	283,507	243,059	40,020	3,996,358
Assessed value for all structures	190,026	109,665	92,524	0	2,396,310
Assessed value missing (dummy)	195,373	0.027	0.163	0.0	1.0
Age (years)	195,373	29.724	27.171	0.0	343.0
Interior square footage (sq ft)	189,564	1,431.779	813.593	0.0	13,940.0
Interior square footage missing (dummy)	195,373	0.030	0.170	0.0	1.0
Parcel size (acres)	195,373	0.575	2.691	0.0	100.0
Townhome (dummy)	195,373	0.201	0.401	0.0	1.0
Basement (dummy)	195,373	0.464	0.499	0.0	1.0
Total # Baths	195,373	1.447	0.945	0.0	20.5
Attached garage (dummy)	195,373	0.245	0.430	0.0	1.0
pool (dummy)	195,373	0.009	0.097	0.0	1.0
pier (dummy)	195,373	0.008	0.089	0.0	1.0
Air conditioning (dummy)	195,373	0.576	0.494	0.0	1.0

3.B. Local Spatial Characteristics

Variable	Obs	Mean	Std. Dev.	Min	Max
Located in high-density residential area (dummy)	195,373	0.201	0.401	0.0	1.0
Located in medium-density residential area (dummy)	195,373	0.517	0.500	0.0	1.0
Located in forest area (dummy)	195,373	0.077	0.266	0.0	1.0
Distance to nearest primary road (meters)	195,373	6,028.952	7,134.341	0.2	46355.2
Bay Depth less than 2 meters (dummy)	195,373	0.981	0.135	0.0	1.0
Located in floodplain (dummy)	195,373	0.053	0.224	0.0	1.0
Distance to nearest urban area (meters)	195,373	23,117.120	13,485.910	100.4	63150.2
Distance to nearest beach (meters)	195,373	9,787.870	9,861.506	7.0	38752.8

3.C. Broader Spatial Characteristics

Variable	Obs	Mean	Std. Dev.	Min	Max
Distance to nearest major city (meters)	195,373	33,143.890	22,116.310	4,776.36	115,250.50
Distance to urban cluster (meters)	195,373	13,919.410	6,505.532	7.82	33,332.02
Distance to wastewater treatment plant (meters)	195,373	8,226.497	6,165.067	41.20	39,470.46
Power plant within 2 miles (dummy)	195,373	0.082	0.274	0.00	1.00
Distance to nearest power plant (if within 2 miles; meters)	16,024	2,130.846	750.966	41.63	3,218.63
% of census block group: high-density residential	195,373	0.110	0.207	0.00	1.00
% of census block group: industrial	195,373	0.015	0.062	0.00	0.84
% of census block group: urban	195,373	0.028	0.062	0.00	0.63

% of census block group: agriculture	195,373	0.105	0.173	0.00	0.85
% of census block group: animal agriculture	195,373	0.001	0.005	0.00	0.17
% of census block group: forest	195,373	0.235	0.188	0.00	0.80
% of census block group: wetland	195,373	0.015	0.041	0.00	0.70
% of census block group: beach	195,373	0.000	0.003	0.00	0.07

a. All dollar estimates converted to USD 2010$ based on the US Bureau of Labor Statistics' Annual US Average "All Urban Consumers – Consumer Price Index (CPI); http://www.bls.gov/cpi/cpid1404.pdf, Table 12 (accessed June 18, 2014).

Table 4. Base Hedonic Regression Results (All 11 counties pooled). Dependent variable: ln(price).

Variables	All Counties Pooled				
	(4.A)	(4.B)	(4.C)	(4.D)[a]	(4.E)[a,b]
waterfront	0.6253***	0.6172***	0.6087***	0.5818***	-
	(0.039)	(0.020)	(0.020)	(0.015)	
× SAV	0.0763**	0.0500**	0.0458**	0.0491***	0.0556***
	(0.028)	(0.020)	(0.019)	(0.019)	(0.017)
water 0-200 m	0.1101***	0.1020***	0.0954***	0.0834***	-
	(0.024)	(0.018)	(0.016)	(0.009)	
× SAV	0.0770**	0.0751***	0.0653***	0.0646***	0.0619***
	(0.033)	(0.029)	(0.023)	(0.019)	(0.019)
water 200-500 m	0.0286**	0.0165	0.0110	-	-
	(0.009)	(0.011)	(0.009)		
× SAV	0.0231	0.0175	0.0173	0.0229*	0.0175
	(0.025)	(0.015)	(0.012)	(0.013)	(0.012)
County Dummies					
x Time Dummies	Yes	Yes	Yes	Yes	Yes
x Home Structure	No	No	No	No	Yes
x Location	No	No	No	No	Yes
Spatial FE	County	CT	BG	BG-BB	BG-BB
(# of FEs)	(11)	(288)	(761)	(1,148)	(1,148)
%Δprice = exp(γ)-1					
waterfront × SAV	0.0793***	0.0513**	0.0469**	0.0503***	0.0572***
	(0.0303)	(0.0211)	(0.0201)	(0.0195)	(0.0178)
water 0-200 m × SAV	0.0800**	0.0780**	0.0675***	0.0667***	0.0638***
	(0.0359)	(0.0312)	(0.0247)	(0.0204)	(0.0199)
water 200-500 m × SAV	0.0234	0.0176	0.0174	0.0231*	0.0177
	(0.0252)	(0.0152)	(0.0122)	(0.0136)	(0.0127)
Obs	195,373	195,373	195,373	195,373	195,373
Adj R^2	0.760	0.671	0.659	0.652	0.670

Note: Clustered standard errors (at the spatial fixed effect level) appear in parentheses below coefficients.

*** $p<0.01$, ** $p<0.05$, * $p<0.1$

a. Coefficient estimates corresponding to *water 200-500 m* not reported because this is the omitted category for each of the block group-bay buffer (BG-BB) fixed effects.

b. Coefficient estimates corresponding to waterfront and proximity buffers not reported because coefficients allowed to vary freely across counties.

Table 5. Additional Block Group-Bay Buffer Fixed Effect Hedonic Regression Results: Homes within 500 meters of Chesapeake Bay Tidal Waters (All 11 counties pooled).[a] Dependent variable: ln(price).

VARIABLES	(5.A)	(5.B)	(5.C)
waterfront × SAV	0.0560***	0.0484***	
	(0.016)	(0.016)	
water 0-200 m × SAV	0.0590***	0.0595***	
	(0.018)	(0.018)	
water 200-500 m × SAV	0.0135	0.0164	
	(0.013)	(0.013)	
waterfront × ln(KD)		-0.0922***	-0.0922***
		(0.025)	(0.024)
water 0-200 m × ln(KD)		-0.0402***	-0.0398***
		(0.013)	(0.013)
water 200-500 m × ln(KD)		-0.0049	-0.0039
		(0.011)	(0.011)
waterfront			
× SAV density 1			0.0571**
			(0.029)
× SAV density 2			0.0517**
			(0.026)
× SAV density 3			0.0679**
			(0.028)
× SAV density 4			0.0312
			(0.020)
water 0-200 m			
× SAV density 1			0.0635**
			(0.025)
× SAV density 2			0.0599***
			(0.022)
× SAV density 3			0.0507**
			(0.020)
× SAV density 4			0.0633**
			(0.026)
water 200-500 m			
× SAV density 1			0.0168
			(0.020)
× SAV density 2			0.0290
			(0.032)
× SAV density 3			0.0048
			(0.015)

× SAV density 4		0.0178
		(0.013)

County Dummies			
x Time Dummies	Yes	Yes	Yes
x Home Structure	Yes	Yes	Yes
x Location	Yes	Yes	Yes
Spatial FE	BG-BB	BG-BB	BG-BB
(# of FEs)	(462)	(461)	(461)
Obs[a]	74,594	74,304	74,206
Adj R^2	0.690	0.690	0.690

Note: Clustered standard errors (at the spatial fixed effect level) appear in parentheses below coefficients.

*** $p<0.01$, ** $p<0.05$, * $p<0.1$

a. Coefficient estimates corresponding to un-interacted waterfront and proximity buffer variables not reported because coefficients allowed to vary freely across counties.

b. Model 5.A includes all 74,594 transactions that are within 500 meters of the waterfront. Model 5.B includes the same set of home transactions, but excludes 290 sales where light attenuation data were missing (138 in Dorchester and 152 in Prince George's Counties). Model 5.C excludes an additional 98 transactions where SAV bed density categories were not matched to the corresponding parcel.

Table 6. Change in Homes with SAV: 2009 Baseline and 185,000 Acre SAV Goal.

	Total	# of Homes w/ SAV=1	
	# of Homes	2009 (baseline)	185k Acre Goal
Waterfront	43,484	5,693	12,286
0-200 meters	40,245	5,043	8,669
Total	83,729	10,736	20,955

Table 7. Total Property Value Differential: 2009 Baseline and 185,000 Acre SAV Goal (million $ USD).[a]

	Approach 1	Approach 2
Waterfront	$ 255	$ 290
	($99 to $410)	($113 to $468)
0-200 meters	$ 71	$ 108
	($28 to $115)	($42 to $174)
Total	$ 326	$ 398
	($127 to $525)	($155 to $642)

a. 95% confidence interval in parentheses.

APPENDIX A. Full Results of Base Hedonic Regressions.

Table A1. Full Results of Base Hedonic Regressions (from Table 4). Dependent variable ln(price).

VARIABLES	(4.A) lnprice	(4.B) lnprice	(4.C) lnprice	(4.D)[a] lnprice
waterfront	0.6253***	0.6172***	0.6087***	0.5818***
	(0.039)	(0.020)	(0.020)	(0.015)
× SAV	0.0763**	0.0500**	0.0458**	0.0491***
	(0.028)	(0.020)	(0.019)	(0.019)
water 0-200 m	0.1101***	0.1020***	0.0954***	0.0834***
	(0.024)	(0.018)	(0.016)	(0.009)
× SAV	0.0770**	0.0751***	0.0653***	0.0646***
	(0.033)	(0.029)	(0.023)	(0.019)
water 200-500 m	0.0286**	0.0165	0.0110	-
	(0.009)	(0.011)	(0.009)	
× SAV	0.0231	0.0175	0.0173	0.0229*
	(0.025)	(0.015)	(0.012)	(0.013)
2nd quarter dummy	0.0142*	0.0158***	0.0167***	0.0176***
	(0.007)	(0.003)	(0.002)	(0.002)
3rd quarter dummy	0.0246***	0.0264***	0.0273***	0.0289***
	(0.006)	(0.003)	(0.003)	(0.003)
4th quarter dummy	0.0375***	0.0413***	0.0432***	0.0448***
	(0.006)	(0.003)	(0.003)	(0.003)
Assessed value for all structures (USD$)	0.0000***	0.0000***	0.0000***	0.0000***
	(0.000)	(0.000)	(0.000)	(0.000)
Assessed value missing (dummy)	0.3227***	0.2867***	0.2671***	0.2582***
	(0.024)	(0.014)	(0.012)	(0.011)
Age (years)	-0.0064***	-0.0057***	-0.0055***	-0.0055***
	(0.001)	(0.000)	(0.000)	(0.000)
Age^2 (years squared)	0.0000***	0.0000***	0.0000***	0.0000***
	(0.000)	(0.000)	(0.000)	(0.000)
Interior square footage (sq ft)	0.0001***	0.0001***	0.0001***	0.0001***
	(0.000)	(0.000)	(0.000)	(0.000)
Interior square footage missing (dummy)	0.1520***	0.1537***	0.1466***	0.1472***
	(0.021)	(0.021)	(0.018)	(0.017)
Parcel size (acres)	0.0178***	0.0184***	0.0192***	0.0189***
	(0.002)	(0.001)	(0.001)	(0.001)
Townhome (dummy)	-0.2802***	-0.2780***	-0.2628***	-0.2643***
	(0.019)	(0.012)	(0.009)	(0.009)
Basement (dummy)	0.0480***	0.0461***	0.0391***	0.0380***

	(0.006)	(0.005)	(0.004)	(0.004)
Total # Baths	-0.0036	-0.0087**	-0.0082***	-0.0085***
	(0.003)	(0.003)	(0.003)	(0.003)
Attached garage (dummy)	0.0852***	0.0561***	0.0467***	0.0436***
	(0.012)	(0.007)	(0.005)	(0.004)
pool (dummy)	0.0426**	0.0350***	0.0355***	0.0291***
	(0.017)	(0.011)	(0.010)	(0.009)
pier (dummy)	0.1242***	0.1251***	0.1320***	0.1305***
	(0.021)	(0.017)	(0.016)	(0.015)
Air conditioning (dummy)	0.0527***	0.0496***	0.0502***	0.0508***
	(0.014)	(0.006)	(0.005)	(0.004)
Located in high-density residential area (dummy)	-0.0941***	-0.1147***	-0.1170***	-0.1088***
	(0.011)	(0.018)	(0.017)	(0.013)
Located in medium-density residential area (dummy)	-0.0835***	-0.0811***	-0.0768***	-0.0723***
	(0.012)	(0.010)	(0.009)	(0.008)
Located in forest area (dummy)	-0.0349***	-0.0297***	-0.0221***	-0.0162**
	(0.011)	(0.010)	(0.008)	(0.008)
Distance to nearest primary road (meters)	-0.0000	0.0000	0.0000	0.0000
	(0.000)	(0.000)	(0.000)	(0.000)
Local Bay Depth less than 2 meters (dummy; 90m2 resolution)	-0.0380	0.0029	0.0089	0.0028
	(0.027)	(0.017)	(0.015)	(0.013)
Located in floodplain (dummy)	-0.0282	0.0080	0.0154	0.0177*
	(0.016)	(0.011)	(0.011)	(0.010)
Distance to nearest urban area (meters)	0.0000***	0.0000	0.0000	0.0000
	(0.000)	(0.000)	(0.000)	(0.000)
Distance to nearest beach (meters)	-0.0000	0.0000	-0.0000	-0.0000
	(0.000)	(0.000)	(0.000)	(0.000)
% of census block group: high-density residential	-0.0934**	-0.0053		
	(0.041)	(0.040)		
% of census block group: industrial	-0.0633	-0.0263		
	(0.095)	(0.064)		
% of census block group: urban open space	-0.1231**	-0.0556		
	(0.047)	(0.083)		
% of census block group: agriculture	-0.0958	-0.0733		
	(0.074)	(0.088)		
% of census block group: animal agriculture	-0.1225	-0.1798		
	(0.736)	(0.480)		
% of census block group: forest	0.0681	0.0735		
	(0.038)	(0.052)		
% of census block group: wetland	-0.0750	-0.0870		
	(0.173)	(0.103)		

% of census block group: beach	0.0182	1.0441		
	(0.319)	(0.634)		
Distance to wastewater treatment (meters)	-0.0000	0.0000		
	(0.000)	(0.000)		
Distance to nearest major city (meters)	0.0000	-0.0000		
	(0.000)	(0.000)		
Distance to urban cluster (meters)	0.0000	-0.0000		
	(0.000)	(0.000)		
Power plant within 2 miles (dummy)	-0.1463**	-0.0563		
	(0.058)	(0.037)		
Distance to nearest power plant (if within 2 miles; meters)	0.0001**	0.0000		
	(0.000)	(0.000)		
Constant	11.7946***	11.9099***	11.8597***	11.8525***
	(0.140)	(0.129)	(0.118)	(0.125)
County Dummies				
x Time Dummies	Yes	Yes	Yes	Yes
x Home Structure	No	No	No	No
x Location	No	No	No	No
Spatial FE	County	CT	BG	BG-BB
(# of FEs)	(11)	(288)	(761)	(1,148)
Observations	195,373	195,373	195,373	195,373
R-squared	0.760	0.671	0.659	0.652

Note: Clustered standard errors (at the spatial fixed effect level) appear in parentheses below coefficients.

*** $p<0.01$, ** $p<0.05$, * $p<0.1$

a. Coefficient estimates corresponding to *water 200-500 m* not reported because this is the omitted category for each of the block group-bay buffer (BG-BB) fixed effects.

APPENDIX B. County Specific Hedonic Regression Results.

Table B.1. County Specific Block Group-Bay Buffer (BG-BB) Fixed Effect Hedonic Regression Results: Set 1 of 2.[a] Dependent variable ln(price).

VARIABLES	(6.A) Anne Arundel	(6.B) Baltimore	(6.C) Cecil	(6.D) Charles	(6.E) Dorchester	(6.F) Harford
waterfront	0.6814***	0.5375***	0.6734***	0.3983***	0.7205***	0.6026***
	(0.029)	(0.050)	(0.095)	(0.124)	(0.073)	(0.084)
× SAV	0.0707	0.0371	-0.0298	-0.0703**	0.0562	0.1328***
	(0.047)	(0.034)	(0.049)	(0.031)	(0.040)	(0.045)
water 0-200 m	0.1237***	0.0518**	0.0800	0.0073	0.0927	0.0144
	(0.020)	(0.025)	(0.092)	(0.090)	(0.056)	(0.060)
× SAV	0.1365***	0.0029	0.0095	0.1032**	-0.0911*	0.0600
	(0.052)	(0.018)	(0.043)	(0.043)	(0.048)	(0.042)
water 200-500 m						
× SAV	0.0664**	-0.0071	0.0140	0.0635***	0.0904	-0.0197
	(0.028)	(0.015)	(0.031)	(0.023)	(0.070)	(0.036)
waterfront × ln(KD)	-0.1007*	-0.1321***	-0.0158	-0.0327	-0.0716	0.0122
	(0.052)	(0.044)	(0.068)	(0.091)	(0.081)	(0.063)
water 0-200 m × ln(KD)	-0.0053	-0.0272	-0.0085	0.0077	0.0221	0.0728*
	(0.023)	(0.018)	(0.037)	(0.065)	(0.056)	(0.041)
water 200-500 m × ln(KD)	0.0665***	0.0034	-0.0496	-0.0434	0.0164	0.0621**
	(0.019)	(0.019)	(0.042)	(0.028)	(0.045)	(0.026)
Time Dummies	Yes	Yes	Yes	Yes	Yes	Yes
Home Structure	Yes	Yes	Yes	Yes	Yes	Yes
Location	Yes	Yes	Yes	Yes	Yes	Yes
Spatial FE	BG-BB	BG-BB	BG-BB	BG-BB	BG-BB	BG-BB
(# of FEs)	(354)	(202)	(61)	(44)	(47)	(76)
Obs	73,521	33,864	10,710	5,211	4,135	17,339
Adj R²	0.643	0.608	0.692	0.725	0.739	0.766

Note: Clustered standard errors (at the spatial fixed effect level) appear in parentheses below coefficients.*** $p<0.01$, ** $p<0.05$, * $p<0.1$

a. Regressions estimated using all transactions in each county that are within 4 km of the Chesapeake Bay or its tidal tributaries.

Table B.2. County Specific Block Group-Bay Buffer (BG-BB) Fixed Effect Hedonic Regression Results: Set 2 of 2.[a] Dependent variable ln(price).

VARIABLES	(7.A) Kent	(7.B) Prince George's	(7.C) Queens	(7.D) St. Mary's	(7.E) Talbot
waterfront	0.9268***	0.0945	0.5371***	0.4711***	0.7842***
	(0.076)	(0.112)	(0.031)	(0.056)	(0.042)
× SAV	0.1053	0.3826**	0.0843**	0.1498*	0.0174
	(0.064)	(0.148)	(0.040)	(0.075)	(0.027)
water 0-200 m	0.2342***	0.0364	0.1332***	0.0150	0.2157***
	(0.071)	(0.038)	(0.032)	(0.027)	(0.046)
× SAV	-0.0182	0.0473*	0.0603	-0.0376	0.1104
	(0.033)	(0.026)	(0.044)	(0.043)	(0.076)
water 200-500 m					
× SAV	-0.0543	-0.0122	-0.0062	0.0376	0.0124
	(0.044)	(0.036)	(0.035)	(0.039)	(0.050)
waterfront × ln(KD)	-0.1835*	-0.0785	0.0219	-0.0254	-0.0612
	(0.094)	(0.068)	(0.053)	(0.063)	(0.051)
water 0-200 m × ln(KD)	-0.0132	-0.0416	-0.1203***	-0.0173	0.0212
	(0.057)	(0.036)	(0.041)	(0.051)	(0.093)
water 200-500 m × ln(KD)	0.0152	-0.0002	-0.0743*	-0.0583	0.1549***
	(0.065)	(0.021)	(0.040)	(0.037)	(0.055)
Time Dummies	Yes	Yes	Yes	Yes	Yes
Home Structure	Yes	Yes	Yes	Yes	Yes
Location	Yes	Yes	Yes	Yes	Yes
Spatial FE	BG-BB	BG-BB	BG-BB	BG-BB	BG-BB
(# of FEs)	(37)	(166)	(45)	(73)	(42)
Obs	3,218	23,379	8,369	5,873	8,153
Adj R^2	0.745	0.614	0.774	0.730	0.704

Note: Clustered standard errors (at the spatial fixed effect level) appear in parentheses below coefficients.

*** $p<0.01$, ** $p<0.05$, * $p<0.1$

a.   Regressions estimated using all transactions in each county that are within 4 km of the Chesapeake Bay or its tidal tributaries.

www.ingramcontent.com/pod-product-compliance
Lightning Source LLC
Chambersburg PA
CBHW080612180526
45168CB00007B/2883